ANIMAL
ARCHITECTS

BOOKS FOR WORLD EXPLORERS
NATIONAL GEOGRAPHIC SOCIETY

CONTENTS

COVER: *Did this young common dormouse in England stop by a nest to visit a bird? No, the mother of the dormouse built the nest. Many animals other than birds build nests for their young.*
OWEN NEWMAN/NATURE PHOTOGRAPHERS LTD.

TITLE PAGE: *Dew glistens on the silken web of a shamrock spider in Wisconsin. Sticky threads will catch insects—the spider's food. Other, dry threads let the spider walk on the web without getting stuck.*
S. J. KRASEMANN/NATIONAL AUDUBON SOCIETY COLLECTION, PR

2

Copyright © 1987
National Geographic Society
Library of Congress CIP data 104

A female mute swan sits on her eggs while the male swims nearby. To build a nest, both birds pile up a huge mound of plants. These swans in Europe use their bills to collect materials. The female uses her belly to shape the top of the mound to hold her eggs.

WINFRIED WISNIEWSKI/FRANK LANE PICTURE AGENCY

INTRODUCTION: WHY ANIMALS BUILD

Animals build structures for the same reasons people do. But there is a major difference. Human architects invent new styles and they choose from among many kinds of building materials. Animal architects build mostly by instinct. They use the same methods and styles, and the same kinds of materials, their ancestors used. On these two pages you'll find out *why* animals build. Then read on to see *how* they go about it.

S. ROBERTS/ARDEA LONDON

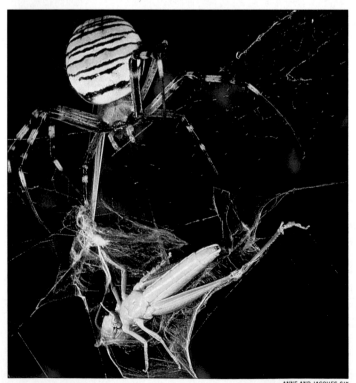

△ **TO AVOID ENEMIES**
Columbian ground squirrels peer from an entrance to their burrow, in western Canada. The burrow is a maze of tunnels that help protect them from predators.

◁ **TO CATCH FOOD**
An Argiope spider has caught a grasshopper in its web, in Europe. When it comes to animals that build structures for capturing prey, you probably think of spiders. Other animals, however, such as ant lions, also build traps to catch meals.

ANNE AND JACQUES SIX

▽ TO STORE FOOD
Several kinds of insects, such as bees and wasps, build structures for storing food. Honeybees build containers of wax to hold honey and pollen that they and their young will eat.

◁ TO ESCAPE HEAT AND COLD
Animals that live in harsh climates may build shelters that help control their environment. In summer, the desert tortoise digs shallow burrows to escape the heat. In winter, a deeper burrow protects the tortoise from the cold.

△ TO RAISE YOUNG
Many animals build nests for the purpose of raising young. A nursery nest may be as simple as a hole in the ground. It may be as complex as a temperature-controlled mound. These young barn swallows stay snug in a nest made of mud, straw, and feathers.

1 Mound Builders

A beaver's long, sharp front teeth strip bark from a small floating log. The beaver will eat the bark, then use the log in building a mound for its home, or a dam to block a stream. In this chapter, read more about the beaver and about other animal engineers that build mounds.

ALAN D. CAREY

TOM DAVIES

Beavers: Work in Progress

When walking in the woods, keep a sharp lookout, especially near streams or ponds. You may notice signs that an animal architect has been at work. The first sign might be a tree stump surrounded by wood chips. If the top of the stump is neatly chiseled, you can be sure that a beaver's big front teeth have been busy there.

If there is a pond, examine it. Is there a large mound of sticks and mud in it? Does the pond spill over a dam of sticks, plants, and mud? The mound and the dam will be sure signs that beavers have been around.

Meet this remarkable animal architect. The beaver, like the mouse, squirrel, and porcupine, is a rodent. It uses its four front teeth—two on top and two below—to gnaw wood. The teeth have a hard orange coating that keeps them from chipping. Like other rodents' teeth, a beaver's front teeth continue to grow throughout its life. Frequent gnawing constantly wears them down.

Unlike most rodents, the beaver has a body designed for life as an architect in the water. When a beaver gnaws sticks underwater, it presses together protective flaps of skin behind its front teeth. This helps keep wood chips and water out of its throat. Clear membranes cover and protect a beaver's eyes underwater, and they enable the beaver to see while swimming. The beaver's ears and nose can pinch shut, keeping out water. Webbed hind feet act as swim fins, helping the animal propel itself. The

◁ A beaver dam holds back a stream, causing a pond to form. In the pond, the beaver builds its dome-shaped mound, called a lodge. The lodge has more than one entrance. Each one lies underwater.

TOM AND PAT LEESON

△ To chew bark or to gnaw wood, a beaver stands on its hind feet and balances with its flat tail. Resting its upper teeth on the wood, the beaver chisels with its lower teeth. It can gnaw through a log the thickness of a rolling pin in just 30 seconds.

KENT AND DONNA DANNEN

△ A hiker examines a tree cut almost all the way through by a beaver. Once a tree is down, the beaver cuts the branches into pieces. A tree this size provides many branches for the beaver's dam or lodge, and plenty of bark for the beaver to eat.

beaver uses its flat, broad, leathery tail to help in steering.

On land, a beaver is slow moving and unprotected from predators. But in water, it can escape most predators by swimming quickly away. To make a safe watery environment for itself, the beaver changes its natural surroundings. A family of beavers can dam a shallow stream and create a deep pond that may cover a small valley.

In winter, a shallow pond could freeze solid. Beavers must make their dam high enough—and the pond deep enough—so that the water will not freeze all the way to the bottom. Water backed up behind the dam often lets beavers swim safely to where there are trees to cut. After felling trees, beavers float the branches to their lodge or dam, letting the water carry the weight of the wood.

To make a dam, a beaver first shoves strong sticks

△ *Jamming in a branch, a beaver reinforces its dam of sticks, plants, and mud. A large dam may stretch more than 300 feet (91 m)* across a valley floor. A dam must be maintained constantly.*

*METRIC FIGURES IN THIS BOOK HAVE BEEN ROUNDED OFF.

▽ *A beaver tows a fresh willow branch toward its lodge. If it does not eat the leaves and bark right away, it will anchor the branch underwater, near its home, as part of its winter food supply.*

ALAN D. CAREY

Making home improvements, a beaver checks the ▷ outside of its lodge. The thick mud-and-wood mound walls keep coyotes and other predators out—and help keep the inside warm and dry.

into the streambed. Then it pushes twigs into the gaps and piles on heavier sticks. Long poles anchor the dam to large growing trees or to boulders on the shore. To make the structure watertight, the beaver plasters mud onto the upstream side of the dam. Low spots, called spillways, allow high water to overflow.

In the pond, the beaver builds its lodge. Only part of the lodge shows above water. At water level, it may be 15 feet ($4\frac{1}{2}$ m) across. If you examine a beaver lodge from the shore, you won't see a way into it. All entrances lie underwater. That makes entry easy for the beaver but nearly impossible for predators. The entrances lead up to a chamber above the surface of the water. The walls may be 3 feet (1 m) thick, making the lodge a secure fortress.

To ventilate the chamber, the beaver leaves a

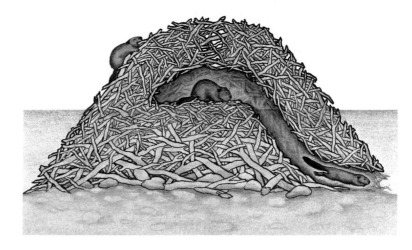

△ *To leave its lodge, a beaver must swim out an underwater tunnel. A lodge has two or more tunnels. Because each one leads underwater, it would be nearly impossible for a wolf or other enemy to enter the lodge. For fresh air, beavers leave air spaces among the sticks in the ceiling.*

column, or chimney, of loosely packed sticks at the top. On cold winter days, when the lodge may hold as many as nine family members, the beavers' warm breath may rise from the chimney in a wispy cloud.

When beavers move into a wooded area, their work may soon destroy most of the trees and flood the land. It may seem like total destruction. But scientists have discovered that beaver engineering often has important benefits in nature. The beaver pond creates a home for fish, water birds, and other animals. The dam prevents soil from washing downstream. The rich soil that builds up soon supports a variety of plants. Eventually, many years after the beavers leave, the land is likely to support a healthy new forest. In the long run, many kinds of plants and animals benefit from the beavers' architecture.

A beaver lodge looks cold and unused in its frozen ▷ *pond, but a family of beavers lives inside, snug and dry. The pond ice presents no problem. By eating branches they have stored underwater, the beavers don't have to leave the pond to find food.*

▽ *Carrying a stick, a beaver swims to one of the entrances to its lodge. A beaver rarely remains underwater longer than 2 minutes, but it can stay under 15 minutes before surfacing to breathe.*

JIM BRANDENBURG

On the dry platform inside her lodge, a female ▷ *beaver crouches near her young, called kits. Shredded wood bedding covers the platform floor. The logs in the lodge show the tooth marks of the beavers that cut them to size.*

ART WOLFE (ABOVE AND OPPOSITE)

Warming Up on the Mound

In the spring, something special starts to happen in wetlands in a few scattered areas of the western United States and western Canada. Trumpeter swans begin to nest. A female swan chooses a nest site. It's usually on a small island in a lake, far from people and surrounded by marsh plants. The swan may repair her nest from a previous year or start building a new one. Often, however, she takes over an abandoned muskrat home—a mound of mud and plants that looks like a miniature beaver lodge.

To build a nest, the female, called a pen, collects vegetation. She scoops it up with her bill and shovels it into a mound. The male, called a cob, also uproots vegetation. He places it at the base of the nest for the female.

The finished nest may measure 5 feet (1½ m) across. Its top stands well above the water, protecting the eggs if the water should rise. The height of the nest also provides the female with a good place from which to spot danger.

After laying her eggs in the nest, the pen sits on them. Heat from her body warms the eggs. When she leaves the nest to eat, she may pull some of the nesting materials over the eggs to trap the heat. The cob stands guard to protect the nest.

Very young swans are covered with fuzzy coats of down, but they cannot maintain their own body temperature on cold days and nights. They need their mother's heat to warm them. So, after the young hatch, the swan family continues to use the nest.

As the young grow, the downy coats get thicker and feathers begin to appear. The young no longer need to be brooded, or kept warm and protected, in the nest. The swans do remain in the area, however, for the rest of the summer. In the fall, they migrate south for the winter.

◁ *Alert to danger, a female trumpeter swan sits on her nest incubating her eggs. She made the mound by piling up materials such as reeds and mosses with her bill. Trumpeter swans use a mound only to raise their young.*

▽ *A trumpeter swan's nest cradles a clutch of eggs. The female usually warms them beneath her belly. When she leaves the nest to eat, she may cover the eggs with nest materials to keep them warm.*

ANIMALS ANIMALS/HARRY ENGELS

△ *Young swans, called cygnets (SIG-nuts), spend a lot of time on the mound, where the female protects them and warms them with her body. Despite this need for care, the cygnets can walk, swim, and find their own food just a day or two after hatching.*

15

It's All a Matter of Degree

When it comes to climate control, the mallee fowl, a bird of Australia, is a master. The male spends about 11 months of every year building and regulating the temperature of the pair's nest.

The process starts in the autumn. The male and female dig a large pit about 3 feet (1 m) deep. In winter, they pile leaves and twigs in the pit. After each rain, the birds cover the vegetation with sand, sealing in the moisture. Eventually, the male scoops out an egg chamber in the center of the leaves and twigs. He fills it with leaves and sand, and then heaps sand in a mound on top.

When moist vegetation decays, it produces heat. So, as the wet leaves and twigs decay, the mound warms up. The male frequently thrusts his bill into the mound, checking the temperature. Four months may pass before the mound reaches the right temperature—about 92°F (33°C). Then, over a period of many days, the female lays eggs in the nest chamber. After each egg is laid, the male covers it with sand. He continually tests the nest temperature with his bill, and adjusts the insulating layer of sand to raise or lower the temperature.

After about seven weeks, the first chick hatches. It struggles up through the sand, and disappears into the brush. From the moment it hatches, fully feathered, it survives on its own. The male tends the mound until all the chicks have made their way to the surface and run off.

△ *A park ranger inspects the giant nest of a mallee fowl, in Australia. The nest is a natural incubator for the mallee fowl's eggs. The male spends about 11 months of each year building the nest and adjusting its temperature.*

△ A female mallee fowl lays an egg in the center of the mound while the male stands by, ready to cover it. The female began laying eggs when the nest reached the right temperature, some four months after it was built. Now she visits the nest every few days until she has laid from half a dozen to more than 30 eggs. Unlike most birds, she will never warm the eggs with her body. Decaying leaves and twigs in the nest produce heat, warming the eggs after her mate buries them there.

After testing the nest temperature with his bill, the ▷ male mallee fowl adjusts the temperature by adding or removing a layer of insulating sand. In summer, he kicks sand onto the nest to shield it from the midday sun. As the day cools, he scratches sand away to allow the sun to help heat the nest—and the eggs buried inside.

Alligator Incubator

Alligators probably aren't the animals that first come to mind when you think of nest builders. But American alligators commonly build large nest mounds that they use for incubating their eggs.

The female alligator is the sole architect of the mound. She usually chooses a spot at the water's edge. Using her sharp teeth, she bites off bits of plant material and shovels the material with her snout to form a mound. At the top, she forms a hollow place for the eggs. Once she lays the eggs, she covers them with vegetation.

This mound of soggy vegetation does just what the nest of the mallee fowl does. It decays, and as it decays, it heats up. The female alligator doesn't work as hard as the male mallee fowl, however. Most of the time, she simply waits near the nest until the eggs are ready to hatch.

When the young begin to hatch, the mother hears them and opens up the mound. The tiny alligator hatchlings will stay near the nest for several weeks while their mother protects them from hungry predators—including other adult alligators.

◁ *An American alligator defends her grassy nest mound in the Okefenokee Swamp, in Georgia. Inside are about 30 eggs, kept warm by the decaying vegetation. The female remains nearby most of the time during the nine weeks before the eggs hatch.*

△ *An alligator wriggles from its egg. Earlier, it grunted as it began to hatch inside the nest mound. Hearing it, the mother used her forefeet to dig into the mound and uncover the egg. The female may carry her young in her mouth to the water nearby.*

19

OBERLE/PITCH

Termite Towers

The world's tallest animal mounds are built by tiny creatures: termites. These insects may build a termite sky-scraper 20 feet (6 m) tall. Their building material, usually soil mixed with their saliva, becomes so hard that some people use it in building mud-and-stick houses.

Unlike the nest mound of the American alligator, which provides warmth to the eggs inside, a termite mound keeps its dwellers cool. The mound, called a ter-mitary (TER-muh-tair-ee), has thick, hard walls that seal in moisture and keep out heat. Inside, a system of chan-nels and ducts circulates air through the mound. These passages run along areas in the termitary walls that are po-rous, or have tiny holes. The pores filter in fresh air for the termites to breathe, and allow stale air to escape.

Within some termite mounds, several types of ter-mites make up a group, called a colony. A single queen produces eggs that are fertilized by a single king. Worker termites take care of the king and queen and tend the eggs. The workers also build and repair the mound, and

◁ *In a hot, dry region in East Africa, a termite mound towers over a visitor. Rising almost 15 feet (4¹/₂ m), the mound helps protect millions of termites from predators, and it keeps them cool.*

GLENN D. PRESTWICH

△ *Steven Carroll, of New York State, examines an opened termite nest in West Africa. He holds a fungus comb, a part of the nest where termites raise mushrooms, a kind of fungus, for food.*

they provide food. Soldier termites defend the colony.

At the base of the termitary are the living and working quarters. A room called the royal chamber houses the queen and king. In other chambers, the workers store food and care for the eggs and the young termites, called nymphs. In some colonies, the workers tend gardens where they grow tiny mushrooms, a kind of fungus.

Termites leave a mound by underground tunnels. The tunnels lead outward and branch into a network of passages that open to the outside. The insects make trips at night to collect leaves, twigs, seeds, and other food.

Some termites in desert areas dig tunnels straight down, more than 125 feet (38 m), to underground water. These wells supply the termitary dwellers with water.

Worker and soldier termites may live for a year or two, and kings and queens for more than ten years. But a termitary may last for close to a century. When a queen or a king dies, another termite takes its place. Generations of termites may inhabit the cool, moist mound.

△ *Australian termites build huge mounds. Inside, a system of channels and air ducts ventilates and cools the nest chambers. Each mound houses millions of termites. Without ventilation and cooling, the termites would die within hours.*

▽ *West African termites have capped their mound with umbrellas of mud that protect the nest from heavy rains. The termites change their architecture to suit their surroundings. In dry areas, they leave off the umbrellas.*

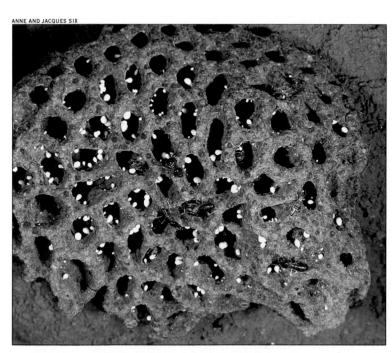

△ *Termites tend the fungus comb inside their mud tower. The comb, made of termite droppings, provides nourishment for the growing fungus (the white masses). The termites feed on both the fungus and the comb.*

Anthill Architects

If you were to stand beside the nest mound of a European red wood ant, you might get quite a surprise. The mound can stand as tall as 6 feet (2 m). The nest may extend underground just as far as it rises above the ground. In the nest live several hundred thousand red wood ants.

An ant colony consists of queens, workers, soldiers, and young. Unlike a termite mound, with one queen, an ant mound may have many queens. When a nest becomes overcrowded, several queens and workers set out to build a new one. They often start with a decayed stump, drilling tunnels and chambers. They dig into the ground below, and add material on top. Soon the stump disappears under a mound of twigs and pine needles.

Like the nest mounds of the mallee fowl and the American alligator, the anthills of European red wood ants are filled with vegetation. Why don't anthills decay on the inside, as do the nests of the mallee fowl and the alligator? In fact, anthills do decay, but only when abandoned. Occupied anthills never decay, because the worker ants constantly carry material from the moist inside of the mounds to the outside. This turnover allows damp material to dry in the sun and keeps it from decaying. In one month, an entire mound may be recycled.

How do the ants stay warm during cold European winters? They rely on the sun. The large, tall nest catches more of the sun's warming rays than a small, low nest would. The sun's heat, combined with the heat of thousands of ant bodies, keeps the colony warm all winter.

During the summer, the ants are fierce predators. A colony may eat huge numbers of insect larvae, or young, in a day. This behavior makes the wood ants perfect pest-control agents. Many countries have laws that protect the mounds of red wood ants. Sometimes, mounds are carefully moved by truck to forests that don't have any.

HANS PFLETSCHINGER/PETER ARNOLD, INC.

△ *European red wood ants scurry home with a pine needle and other building material for their anthill in a forest in northern Europe. They carry some material to the top of the mound. More is dropped on the sides, creating a dome-shaped mound.*

22

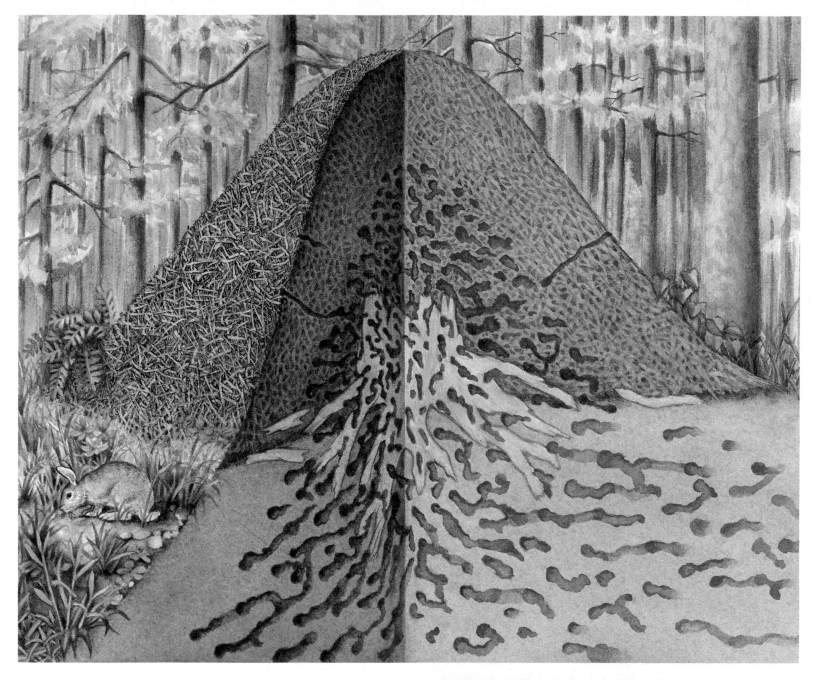

△ The mound of a European red wood ant is larger than any ant mound you'd see in North America. It may stand as tall as 6 feet (2 m). Thousands of ants live in it. They collect tiny bits of the forest around them and pile the pieces on a stump. The huge mound covers and protects a series of passageways and chambers where the ants live, breed, and raise their young.

European red wood ants scurry over the twigs, pine ▷ needles, and bark of their mound. They constantly move damp materials from the inside to the outside of the mound, preventing the mound from decaying.

E. S. ROSS

2 Weavers And Platform Builders

Carrying fresh nesting material, a male red-headed weaver arrives at his partially completed nest, in Africa. He will secure the new strip of vegetation by weaving it into the ring of plant stems. This chapter tells about a variety of animals that build structures by weaving or piling materials.

Busy as a Weaver

Yes, indeed, weaverbirds do weave! The males, the ones that weave the nests, practice as youngsters. By the time they are old enough to mate, they're expert weavers.

To collect material for the nest, a male masked weaverbird flies to the base of a long green blade of grass. Pinching one side of it in his bill, he flies away, ripping off a narrow strip of the blade. Although his nest will dry to a golden brown, the weaverbird builds with green grass. This assures him of flexible construction materials.

The masked weaverbird first weaves a ring of grass. Then, perching in it, he weaves a hollow ball around himself. This forms the main chamber. Then he adds an entry chamber below. When he's finished, the bird hangs under the nest and calls to females. A female may fly over to inspect the nest. If she is satisfied with it, she lines the nest with soft grasses and moves in. If she's dissatisfied, the male picks apart his work and starts all over again.

△ To anchor a blade of grass to a twig, a masked weaverbird uses its feet and bill. It holds one end of the blade with a foot while using its bill to wind, weave, and knot the other end. The bird often uses grass to hold together two small twigs to make a stronger support for its nest. *

JEN & DES BARTLETT/SURVIVAL ANGLIA (BOTH)

△ A masked weaverbird in southwestern Africa selects the end of a thin forked branch for his nest site. The forked branch serves as the foundation for the basketlike nest the weaver is just beginning.

△ The weaverbird has woven strips of grass to make a sturdy ring. Sitting in the ring, he will weave a nest chamber around himself.

*UPPER ART: ADAPTED FROM ANIMAL ARCHITECTURE, COPYRIGHT © 1974 BY KARL VON FRISCH AND OTTO VON FRISCH; DRAWING COPYRIGHT © 1974 BY TURID HOLLDOBLER. REPRINTED BY PERMISSION OF HARCOURT BRACE JOVANOVICH, INC. LOWER ART: ADAPTED FROM N. E. COLLIAS

Nests of masked weaverbirds hang from a branch like fruit. The roofs keep out rain and sun. The nests open at the bottom, making it hard for tree snakes to enter. Only males weave nests. If a female doesn't approve of a nest, the male builds another.

Haystack Homes

If you travel in parts of southern Africa, you will probably see the nests of birds called sociable weaverbirds. These straw homes may stretch 15 feet (4½ m) across and hold as many as 125 nesting pairs. Used year after year, and added onto by new nesting pairs, these homes are the largest bird nests you'll ever see in a tree.

How is such a large nest formed? Several pairs of birds start building together around a sturdy branch on a big tree. Thorn trees, which are common, are popular nest sites, but sociable weaverbirds aren't terribly choosy. They sometimes build their homes on telephone poles.

The birds collect thatch, or coarse grass, and weave it into an umbrella shape, curved on top. Into the underside the birds weave individual nest chambers. Each of the many chambers is lined with bits of grass. The walls between nest chambers may be flimsy, but all the chambers are protected from rain by the waterproof roof.

In their thatch-roof "cottage," the birds are safe from enemies. Occasionally, however, they find themselves with uninvited neighbors. Small parrots, pygmy falcons, and other kinds of birds move into unoccupied nest chambers to raise their own young.

As a result of their hard work, the weaverbirds sometimes end up with a major problem. After many years, a nest may become so large and heavy that its supporting branch breaks under the weight. If this happens, the nest tumbles to the ground. The sociable weaverbirds move to another tree and start a new colony.

◁ A treetop haystack? No, it's a group nest built by birds called sociable weaverbirds, in southern Africa. As many as 125 pairs may nest in it. The birds share a single thatch roof, but each pair has its own nesting chamber with a separate entrance.

△ Sociable weaverbirds return to their nest. "Sociable" suggests togetherness. These birds are well named: They fly, feed, and nest in groups.

△ Monk parakeets nest very much the way sociable weaverbirds do. The parakeets build a bulky nest of twigs. Twenty or so birds may move in. Although monk parakeets are native to South America, they can be found in Florida. Brought to Florida as pets, some escaped and now live there in the wild.

Sticks and Cones

When you think of busy builders, you might think of beavers or bees, but . . . red squirrels? Yet the red squirrel is such an energetic builder that it may make two types of homes—in addition to a home it may simply take over.

A red squirrel that finds a cavity in a tree will use it as a home. Old holes made by woodpeckers are common sites for squirrel homes. When suitable holes are not available, however, a squirrel builds a home called a dray.

The red squirrel builds a sturdy dray in the fork of a tree, close to the trunk. There, the dray won't be shaken loose by strong winds. The animal breaks off several branches and props them in the fork. On top of these it piles twigs and leaves. In the center it forms a round hollow with an entrance on one side.

During winter, the squirrel lives in its tree hole or in its dray. It may stay inside for days at a time in bad weather. In summer, it may use either of these homes, or it may build a loose, temporary dray of twigs and leaves. It uses the loose dray for resting or for hiding from enemies.

If you hike in woods where red squirrels live, you may find large piles left behind by the squirrels. The piles, called middens, form by chance as squirrels sit in favorite spots and break open nuts, seeds, and cones. As squirrels split open their food, nutshells, seed husks, and cone scales fall to the ground. Over time, the bits pile up in sizable heaps. Squirrels use these middens, among other sites, to store unripened cones and other food for the winter. When a squirrel needs a meal in winter, it digs down to the midden, sometimes through many feet of snow, and pulls out a snack.

◁ A bundle of vegetation serves as a squirrel's sturdy nest. A red squirrel may build a nest like this one for warmth in winter and as a nursery. Just before giving birth, the female chases the male away. She raises her young alone. Red squirrels live in forests in North America, often in mountainous areas.

△ What's this pile? Red squirrels sat directly above it, breaking open pinecones. Pieces of the cones dropped, forming the pile. Squirrels store food inside such piles for winter. Burrowing in, they may discover additional food dropped accidentally.

△ Nibbling a seed on a stump, a red squirrel may chatter loudly. Its calls have earned it the nicknames "chickaree," for the sound it makes; and "boomer," for the volume of the sound. The squirrel builds its nest with its mouth and paws.

31

Wraparounds

The little harvest mouse of Europe and parts of Asia builds a tiny, tightly woven nest. You might think the nest belonged to a very small bird.

The mouse weaves the nest around stalks of wheat, oats, or tall grass. Such slender supports would collapse under the weight of a larger creature.

Both males and females build nests, but the nests are used for different purposes. The male harvest mouse builds a rather loose nest and uses it only for sleeping. The female builds a sturdier nest. In it, she bears her young and raises them.

A harvest mouse uses its teeth and delicate front paws to shred blades of grass into long, thin pieces much like those that weaverbirds use. It wraps the pieces around the supporting stalks, weaving a tight hollow ball. It leaves an opening for the entrance on one side.

A female harvest mouse collects thistledown and cattail fluff from nearby plants to make herself a soft, comfortable bed inside the nest. She gives birth to a litter of three to five sightless, hairless babies. The female leaves the nest to feed on grain and other seeds, but returns often to nurse her young. The young stay in the nest until they are ready to leave it for good.

Within two weeks of their birth, the tiny pink baby mice grow into lively miniatures of their parents. As they grow and nose around, they try to tunnel through the walls of the nest. They often tear it to shreds.

Once out on their own, the young mice build nests for sleeping. At first, the inexperienced mice make loose and messy nests. After some time, they learn to build stronger ones.

The female mates again shortly after giving birth. When the first litter grows up and leaves the nest, the female prepares a new nest, where she gives birth to a new litter. She may build several nests in a single season, weaving each with care and skill for her next litter.

ANIMALS ANIMALS/OXFORD SCIENTIFIC FILMS

◁ *Newborn harvest mice—hairless, with eyes not yet open—sleep soundly in their nest. In just two weeks, these sleepy young will become so active that they may tear the nest to pieces. Their hardworking mother may have several litters in one season. She builds a new nest for each litter.*

An adult harvest mouse climbs from its nest in a ▷ field of grain. The nest is wrapped around several stalks that support it 1 to 1½ feet (30 to 46 cm) above the ground. Like some monkeys, the mouse can use its tail to grasp things. This ability makes it an expert climber. The mouse can hold on to grain stalks by its tail and hind legs while using its paws and mouth to weave its nest.

The various bowerbirds of Australia and nearby New Guinea might be called true animal artists. The males construct bowers—walkways of twigs and grasses. They decorate the bowers to attract mates.

To build its bower, the male satin bowerbird, a dark blue bird of Australia, first clears an area about 3 feet (1 m) square. Then he pushes straight twigs into the ground. The avenue created between the twigs usually runs north-south. At one end, the bird lays down a platform of grass and small twigs. The platform will serve as a stage for the decorations the bowerbird collects.

The bird has just begun his work. Now he turns his attention to decoration. He searches the area for blue—or sometimes green—objects, such as flowers, berries, feathers, dead beetles, or anything else of the right color that he can carry. If manufactured objects are available—clothespins or marbles, for example—the bird adds them to his collection in front of the bower.

The satin bowerbird even paints his bower. He crushes blue berries in his bill. Then he adds saliva to the mash and rubs the mixture on the bower walls. A male sometimes raids the bower of another male, making off with the best decorations.

When a female appears near the bower, the male begins to dance. If the female is attracted by the dance and sees an impressive number of decorations, she enters the bower and mates with the male.

The female bowerbird doesn't remain at the bower for long. After she and the male mate, she flies away. She builds a shallow, saucerlike nest a short distance from the bower. There, she lays eggs, incubates them, and raises the young. Back at the bower, the male picks fresh flowers and waits for the next female to come along.

Every year, the satin bowerbird builds a new bower. The golden bowerbird, also of Australia, keeps the same one year after year. Both kinds of birds, however, must make occasional repairs throughout the mating season. The arrivals and departures of the birds, and wind and heavy rain, damage the structures. The birds redecorate the bowers constantly. Inspecting their structures, they remove withered flowers and dried berries, and replace them with fresh ones.

◁ *A dove-size satin bowerbird patrols his U-shaped structure, called a bower. The male builds the structure of twigs to attract a mate. He decorates it with flowers and berries, and perhaps even with pencils and clothespins. This Australian bird usually chooses blue items for decoration.*

One of the smallest bowerbirds—the golden ▷ bowerbird of Australia—builds the largest bower, adding to it year after year. The male finds two saplings about 3 feet (1 m) apart and piles twigs around them. The bower is not a nest. It is designed only to attract females. When a female comes into the area, the male dances in front of his bower. He continues to dance until she mates—or leaves to visit another bower. After mating, a female builds a nest elsewhere and raises the young.

◁ A golden bowerbird decorates his bower with lichens (LYE-kunz), or mosslike plants. Different kinds of bowerbirds use different colors. The golden bowerbird prefers blue-gray or green decorations. The branch he perches on connects the sides of his bower. On the branch, he will display his feathers for a female attracted to the bower.

They build bowerlike nests for mating, but they're not bowerbirds. They're three-spined sticklebacks. These small fish, just 2 inches (5 cm) long, live in coastal waters, rivers, and lakes in the Northern Hemisphere.

When it's time to mate, the male chooses a sandy spot and makes a shallow pit with his mouth. Then, he gathers plants and piles them up. He places the plants on the bottom lengthwise, with gaps between them. The gaps will help ventilate the finished structure when it's used as a nest. When the pile is high enough, the fish coats it with a sticky substance made in his kidneys. This cements the mound together. Next, the male burrows through the mound, creating a tunnel-shaped nest.

Now the male seeks a mate. By the time the nest is built, the male has developed a bright red coloring on the underside of his head and belly, and his back has turned bluish white. These colors attract female sticklebacks.

When a female swims by, her belly swollen with eggs, the male begins a zigzag dance. He may brush the female with his stickles—the spines on his back. If she responds, the male escorts her to the nest. The female goes inside and deposits some eggs. Then she swims away to lay more eggs in other nests. After the male fertilizes the eggs, he tries to attract more females.

When several hundred eggs have been laid, the male begins to tend them. He fans the water over them with his tail. The water moves through the nest, carrying oxygen to the eggs. After the eggs hatch, the male looks after the young for a few weeks. If a baby strays, the male captures it in his mouth and spits it back into the nest.

△ *A male stickleback moves sand aside with his mouth, clearing a construction site. In this shallow pit he will build a nest. Then he will attract females there and tend their eggs.*

▽ *The stickleback collects algae (AL-jee) and other water plants for the nest. When he has made a mound of vegetation, he will add a sticky glue from his kidneys. He then will burrow through the cemented mound, creating a tunnel-shaped nest.*

A male stickleback displays for a ▷
female at the entrance to his nest.
The female hovers over him, her
belly swollen with eggs. His bright
colors are part of his courtship
display. If the female approves of
the male's display, she'll swim
into the nest.

▽ Once the female is inside the nest,
the male nudges her, encouraging
her to lay eggs. After laying a
number of them, she will leave.
The male will fertilize the eggs,
then guard and tend them, and
finally watch over the young.

Nests Ahoy!

They're not exactly houseboats, but the nests of some kinds of birds float!

Horned grebes usually build their floating nest in a sheltered part of a lake, using reeds. They pile the reeds together until the mass is thick enough to bear their weight—and the top of the nest is well above water. Then they line the nest with soft material, such as moss. The finished nest is damp. Scientists think that when a parent incubates the eggs, heat from its body heats the moisture in the nest, helping warm the eggs.

Magpie geese build a more elaborate nest. The male first constructs a platform on wet or marshy ground. He bends over plant stalks with his bill or neck and flattens them with his feet. On the platform, the male cleans and adjusts his feathers, and courts females. If rain raises the water level, the platform floats.

Just before egg laying begins, the male works on the nest again, with one—or sometimes two—mates. They bring uprooted plants and pile them on the platform, leaving a spot in the middle uncovered. There, the female or females lay up to nine eggs. In the heat of day, the birds may not sit on the eggs. They may even stand up and shade the eggs to keep them from getting too hot in the sun. At night, the birds do incubate the eggs.

◁ A horned grebe sits on its nest of cattails, bullrushes, and other reeds. If the water level rises, the nest floats. If the water level drops, the nest drops. The horned grebe lives in North America, Europe, and Asia.

▽ This Australian bird, running ahead of its young, hatched them on a floating nest in a swamp. Scientists aren't sure how to classify the magpie goose. Together, the bird's hooked bill, its partly webbed feet, and its behavior set it apart from other waterfowl. One unusual fact: A male magpie goose often has two mates. All help raise the young.

△ A magpie goose uses its neck to bend down plants in shallow water. Then it tramples them flat. Once a platform is built, uprooted vegetation is added to form a central cup. The females lay eggs in the cup.

40

Have you ever seen the nest of a robin, a blue jay, or a mockingbird? Such nests are tiny compared with the nests some other birds build. Ospreys and white storks, for example, make huge nests. Ospreys often nest in tall trees or on high platforms that people build to encourage these birds to nest. White storks usually nest on the roofs of houses, barns, or churches, but they may also use trees.

Ospreys build their nests mostly of sticks and branches. An osprey usually drops a branch on the nest and then shoves it around with its beak until the branch catches firmly. Ospreys, also called fish hawks, mate for life. Year after year, a pair will add to their nest. They line it with fresh grass and moss, and raise their young.

Storks do not mate for life. They choose new mates each year. In the spring, a male looks for an existing nest, but not necessarily the one he used the year before. If he comes across an occupied nest, he may try to force out the stork already there. Fights often break out between the occupant and the intruder. The winner keeps the nest, and the loser flaps away in search of another nest—or a site for a new one.

Once a male stork has a nest, he attracts a mate. Together, they start repairing the nest. They add new branches and fresh lining material.

People in Europe consider storks to be symbols of good luck, and welcome them to their rooftops. Homeowners lucky enough to be hosts to a nest protect it.

It's an old tradition for people to care for stork nests. Historians have found evidence from the year 1549 showing that a homeowner in Holland paid for repairs to the stork nest on his roof. In 1930—381 years later—storks were still using this same nest.

◁ *An osprey lands on its enormous nest beside its mate. The pair will raise their young here year after year, adding new branches, driftwood, seaweed— even bones and old shoes—each breeding season. Osprey nests eventually become so large that you can see them for miles. Ospreys live near water— along seacoasts, and near lakes, ponds, and slow-moving rivers—wherever they can catch fish.*

Symbols of good luck, a pair of white storks nest ▷ *on a rooftop in Poland. These large birds will migrate to Africa in the winter and return to Europe in the spring to breed. For hundreds of years, Europeans have welcomed storks to the rooftops of their homes. People put up platforms, cartwheels, and even false chimneys to give the birds supports for their nests.*

M. P. KAHL/DRK PHOTO

Tree House

Chimpanzees travel around the green African forest in search of food. As darkness falls, these apes are often far from their previous night's location. The group selects one large tree or several trees near each other. In just a few minutes, each chimp weaves a comfortable nest for itself as high as 80 feet (24 m) above the ground.

If the chimps remain in the area the next day, they may use the same nests that night. Usually, however, each animal builds a new nest each night.

A young chimp sleeps in its mother's nest. Such a nest is larger than normal, and it takes longer to build. The mother often holds the baby in one hand while using her other hand and her feet to build the nest. As the youngster grows up, it practices building platforms, often jumping up and down on them in the trees. By the time it is about 2½ years old, it can build its own nest.

◁ Sprawled out on its day nest of branches and leaves, a chimpanzee snoozes. Chimps use nests only as sleeping platforms. Sometimes they make flimsy day nests for naps. For the night, they build sturdy nests that hold them securely.

△ To make a sleeping platform, a chimp bends long branches over and under the larger forked branches of a tree. It flattens the bed with its feet and adds leaves and twigs for softness.

43

3 Users Of Silk, Saliva, And Foam

Wrapped in the silk threads of its own cocoon, a polyphemus caterpillar will soon develop into a moth. Read on to find out about other creatures that spin silk for homes. You'll also meet animals that build nests from other substances they produce in their bodies.

E.S. ROSS

If you're walking among trees and you see a grayish white silken mass on a branch, it's probably a shelter for tent caterpillars. As many as 300 caterpillars work together to build such a tent. They are the larvae that hatch from the eggs of a tent caterpillar moth.

To make the tent, the caterpillars spin silk, wrapping it around and around a leafy branch. The tent helps protect them from stinkbugs and other enemies. By day, the caterpillars leave the tent to feed on nearby leaves. As they move around among the branches, they lay down trails of silk. At nightfall, they retrace the silken trails and return to the tent.

When the caterpillars are ready to pupate (PYEW-pate), or develop into moths, they leave the tent. Each finds a hidden place and spins itself a cocoon.

Another caterpillar, called a bagworm, constructs a case around itself soon after hatching from its egg. The bagworm finds twigs or leaves in the tree or shrub where it feeds. It weaves these together into a silken case. As the bagworm grows, it adds to this "armor." The animal carries the protective case along with it as it moves around, poking out its head to feed.

When the bagworm is full-grown, it uses silk to anchor the case to a branch or leaf. Sealing the opening with silk, it spins a silk inner case, or cocoon. There, the caterpillar pupates. The adult male develops wings and leaves his cocoon to mate. The adult female never leaves her cocoon. She lays eggs in it. When the eggs hatch, the larvae crawl out of the case and move away, each to make its own tiny new case.

Eastern tent caterpillars swarm in and out of their silken shelter. As many as 300 of them may live in a single tent. When they are ready to develop into moths, they leave the tent and spin individual cocoons.

DWIGHT R. KUHN

△ An Australian bagworm, a kind of caterpillar, adds a twig to the case it has built around its body. It attaches the twig using its own silk. The bagworm will add to the case as its body grows larger. The case may reach 3 inches (8 cm) in length.

Carrying its shelter as it goes, an African bagworm ▷ reaches out of its case to feed. The case, built of twigs from nearby bushes, may be hard for enemies to see. When the caterpillar has reached its full growth, it attaches the shelter to a leaf or branch. Then it spins a cocoon inside the case.

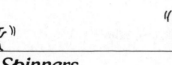

Super Spinners

When you think of spiders spinning silk, you probably think of them making webs to trap insects. But spiders use silk for many purposes. They use it to make trapdoors, diving bells, safety lines, and other devices.

Spider silk is one of the marvels of the natural world. It is produced in silk glands and squeezed out through spinnerets at the rear of a spider's body. Some kinds of spider silk are stronger than steel of the same thickness. Sticky silk stretches—sometimes more than three times its original length. Spiders can recycle silk by eating old or torn webs. Strong and light, spider silk is used by other builders. Some birds use it in building their nests.

Spiders depend on silk from before they hatch until they die. They hatch from eggs a female laid and wrapped in a silken sac. Immediately upon leaving the sac, some kinds of spiders attach a silk safety line to a nearby object. Like rock climbers on a cliff, those spiders continually anchor their safety lines as they travel. If a spider slips, a silk thread stops its fall. As the spiders grow, they continue to

MICHAEL FOGDEN/OXFORD SCIENTIFIC FILMS

△ An Argiope (are-JYE-uh-pea) spider hangs in the lacy center of its web, in Central America. Some scientists think the bright white zigzag pattern alerts birds that might otherwise fly through the spiderweb, destroying it.

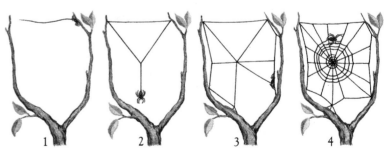

◁ Aided by claws and hairs on the ends of its legs, a garden spider can travel safely across its web. It walks mostly on the dry "spoke" threads, avoiding the sticky circular threads, which trap insects.

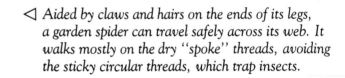

△ 1) An orb weaver starts its web by using the wind to carry a line of silk across a gap. 2) The spider travels across the first line, laying down a second one. Then it moves back to the middle, attaches a hanging thread, and drops to the bottom twig, pulling down the second line. 3) The spider rapidly attaches more of these "spoke" threads. 4) Now, starting in the center, the spider lays down a temporary spiral of dry silk that holds the spokes in place. Later, retracing its steps, the spider rolls up the dry spiral and lays down a sticky one.

ART ADAPTED FROM PAINTING BY PETER BIANCHI UNDER THE DIRECTION OF JOHN A. L. COOKE © NATIONAL GEOGRAPHIC SOCIETY

HANS PFLETSCHINGER/PETER ARNOLD, INC.

△ *A monkey spider guards the funnel-shaped silken entrance to her burrow, in Africa.*
The spider hunts at night using her sense of touch. When an insect stumbles across the
silk lines that attach the funnel to the ground, the spider will feel the vibrations and
rush from her burrow to catch her prey.

49

use safety lines. All of their lives, they keep themselves attached to the world by a silk strand.

Have you ever brushed a spider off your shoulder or arm? Think back. Did the spider tumble to the ground or did it slide down a safety line?

Web-spinning spiders depend on their silk for more than safety and catching insects. They use it to communicate with others of their kind. A male seeking a mate uses his legs to pluck threads on the web of a female. The plucking causes the web to vibrate. If the female accepts the male, she taps out a message to him on the web or she shakes the web.

During courtship, spiders use silk in yet another way. The males of some species present the females with flies—gift wrapped in silk.

△ *A fierce predator, the wolf spider approaches the entrance to her home, in Africa. The silken trapdoor, attached to the burrow by a strong silken hinge, lies open, exposing the silk-lined burrow. When the spider goes inside, it pulls the door shut and lies quietly below. When an insect passes by, the spider senses the vibrations. It throws open the trapdoor and pounces on the prey.*

◁ *A jumping spider in Oklahoma peers out of its dangling home—a bag of silk it spun within a curled leaf. The spider doesn't spin a web. To catch prey, it jumps on insects.*

◁ A female wolf spider in Europe puts the final silken touches on an egg sac. She started spinning the bottom and sides before she laid her eggs. After depositing her eggs, she spun a lid. Now she wraps the egg sac with silk threads.

△ Have you ever turned a cup upside down, pushed it into the water, and noticed how air is trapped inside? Water spiders, native to Europe and Asia, actually build themselves upside-down "cups" underwater and fill them with air. A spider first spins an underwater silken tent, which acts like a cup. The spider attaches the silk to plants. Making several trips to the surface, it swims back down to its tent carrying air bubbles trapped among the hairs on its abdomen and legs. The tent soon swells with air. It becomes a diving bell, or underwater living quarters. From there, the spider attacks prey in the water.

Child Labor

For a long time, people interested in animal architecture were puzzled by the nests of weaver ants. The nests are clumps of living leaves pulled together and held in place with silk. Adult weaver ants, however, cannot make silk. Ant larvae make silk, but the helpless larvae can't crawl around. How was the silk put in place?

A naturalist solved the mystery about a hundred years ago. He cut a small slit in an ant nest. Soon, weaver ants swarmed out to repair the nest. Living chains of ants tugged the separated leaves back together. Other ants scurried forward, carrying larvae. Holding the larvae in their jaws, the worker ants moved back and forth across the slit. As the ants carried the larvae from leaf to leaf, silk threads made by the larvae stuck tight at each place touched by a larva. The naturalist had discovered that the ants used the larvae to "lace" leaves together.

▽ *The weaver ant on the left uses an unusual tool—an ant larva—to attach two leaves together. The leaves will help form a nest. The ant holds the larva in its jaws as the larva produces silk. As the silk comes out, the ant moves the larva back and forth between the leaves. The sticky silk holds the leaves together.*

Weaver ants swarm over their nest of leaves woven ▷ together with silk. They use only living leaves. When the leaves die, the ants make a new nest. Weaver ants live in Africa, Asia, and Australia.

Swift Work

Many of the birds called swifts and swiftlets live up to their names. They are fast and acrobatic.

All of these birds use saliva as a glue in building their nests. Most use twigs and other vegetation as well.

As their name suggests, chimney swifts commonly build nests and raise young inside chimneys. Because these North American swifts breed during the warm months, they are usually safe from fires that people might start in fireplaces. Chimney swifts sometimes nest in hollow trees, in barns and garages, and even in old wells.

At the start of the breeding season, the salivary glands of chimney swifts enlarge. These glands, which make saliva, swell to several times their normal size, giving the birds an enormous supply of saliva. Besides being useful in nest building, the saliva serves another purpose. Little balls of insects stuck together with saliva are fed by the swifts to their young.

Swifts and swiftlets are so well adapted to the air that they rarely land, except to sleep or nest. They catch insects in flight. They even collect nesting material without landing. With their feet or bills, they break off twigs and catch feathers and dry grasses that blow by.

◁ *A chimney swift in Ohio incubates eggs in its nest of twigs and saliva. The swift used saliva to cement the nest to the inside wall of a chimney.*

▽ *To build this nest, in Southeast Asia, a white-bellied swiftlet first set down a gummy base of saliva on a rock wall. The bird built up the cup shape by mixing vegetation with layers of saliva.*

Edible-nest swiftlets, which live in tropical Asia, ▷ *build nests that people value highly. They are made entirely of swiftlet saliva. People collect the nests from high cave walls. Chefs then clean and stew the nests to make bird's-nest soup.*

Air Care

The male Siamese fighting fish is both architect and caretaker. Gulping air and spitting it out again and again, he makes a foamy mass of bubbles for a nest. He anchors the nest at the surface to part of a water plant.

While building his nest, the male attacks any fish that comes near. Once finished, he mates with a female beneath the nest, fertilizing her eggs as she lays them. The male then scoops up the eggs in his mouth and deposits them in the nest.

The male soon drives the female away. After the eggs hatch, the male watches over the hatchlings for a few days. Then the young swim off on their own.

A male Siamese fighting fish swims beneath his nest ▷ of bubbles. As a female lays eggs under the bubbles, the male fertilizes them, then catches them in his mouth and places them in the nest. The bubbles provide shelter for the developing young.

ANNE AND JACQUES SIX (BOTH)

△ *To make the bubbles, the male Siamese fighting fish gulps air. Mixing it with mucus, a sticky fluid in his mouth, helps make the bubbles last.*

56

Foam Home

After a storm during the rainy season, African gray tree frogs come together to mate. These tree-climbing frogs find one another by making squeaky calls. Sometimes, groups gather on a branch that extends over water. There, each female produces a sticky fluid. The frogs kick at the fluid with their powerful hind legs, pumping air bubbles into it. A foamy mass soon accumulates, like lather on a bar of soap. In the foam, each female lays about 150 eggs. The males immediately fertilize them. The females then produce more fluid over the eggs. The frogs beat this into an added layer of foam. When the adult frogs have made a white mass about the size of a volleyball, their job is finished.

The outside of the foamy nest hardens into a crust that keeps the moist contents from drying out. Inside, tadpoles develop from the eggs. As the tadpoles grow, they develop structures called gills that will enable them to take in oxygen underwater.

When the tadpoles are ready to leave the nest—about five to eight days after the eggs were laid—rain usually falls again. The rain makes the foam nest soft and heavy. The nest drops into the water piece by piece, carrying the tadpoles along. Once in the water, the tadpoles complete their development. Slowly growing hind legs, then front legs, and then losing their tails, they turn into miniature versions of their parents.

△ Clinging to a branch over water, African gray tree frogs use their hind legs to whip up fluid produced by the females. As the females lay their eggs in this foamy mass, the males fertilize the eggs. The frogs will then beat the foamy nest into a larger mass.

JASON RUBINSTEEN/WEST STOCK, INC.

△ A nest of African gray tree frogs hangs suspended over water. The outside of the nest hardens into a crusty, protective material. Eggs in the moist interior develop into tadpoles. A nest may be built as high as 50 feet (15 m) above the water.

In a few days, rain softens the foam. The nest ▷ drops in pieces, along with the tadpoles, directly into the water. The young tadpoles shown here have developed enough so that they are ready to swim. They have grown structures called gills, through which they will take in oxygen underwater.

59

MICHAEL FOGDEN/BRUCE COLEMAN INC.

Bunches of Bubbles

Blades of tall grass serve as the foundation for the nest of a spittlebug. The insect's nest looks like a gob of spit. A spittlebug builds the nest while young, when it is called a nymph. The nest protects the nymph's tender skin from the sun, and helps hide it from predators.

How does the nymph make its spit? First, it makes a small pool of liquid. Then it introduces air into the liquid through a tube on the end of its abdomen. The air makes the liquid bubble up into foam.

You may wonder what makes the foam last. The nymph's kidney tubes added a special ingredient to the liquid before the foam was made. The foam is present until the nymph develops into an adult—a process that takes 30 to 100 days, depending on the temperature.

Visit a meadow in the springtime and you might see ▷ a scene such as this—dozens of nests of spit in the tall grass. Each nest houses a young spittlebug, called a nymph.

DWIGHT R. KUHN

△ *Leaving its nymph skin behind, an adult spittlebug climbs from its foamy shelter. The spittlebug developed safely inside the foam nest.*

60

Animals That Build With Mud

4

Feeding a hungry chick, a barn swallow balances by beating its wings. Its cuplike nest of mud and straw, plastered to a sheltered wall, cradles the chicks. In this chapter, you'll read more about swallows and other animals that use mud to build structures of various shapes and sizes.

CHUCK J. LAMPHIEAR/DRK PHOTO

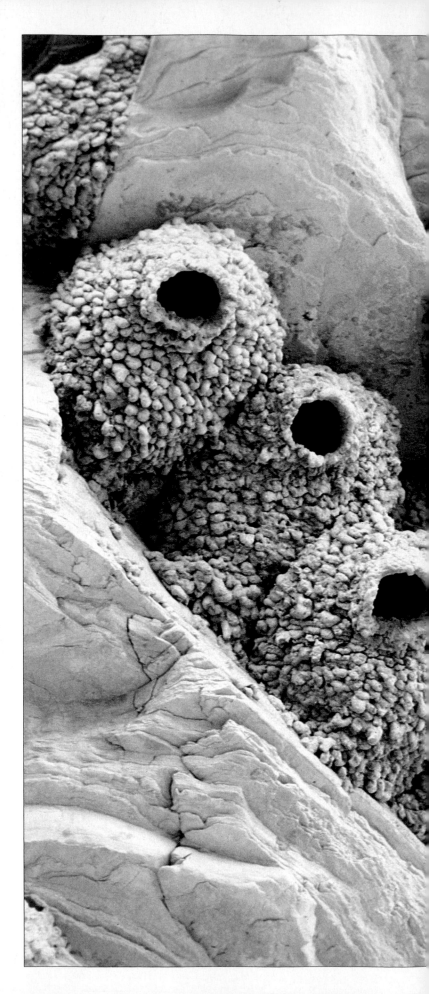

Cliff-hangers

Few other animals build where cliff swallows often do—on the sides of rocky cliffs. In North America, where cliff swallows nest, you may find hundreds of their jug-shaped structures stuck to a single cliff. They also nest in barns and under bridges. Because the nests are so hard for other animals to reach, they provide the swallows with safe places to lay eggs and raise young.

Cliff swallows use their bills to gather mud for their nests. They carry the mud as pellets, which they stack up like bricks to build a nest. First, they lay down a row of pellets about 4 inches (10 cm) below a rock overhang. Then, layer by layer, they add pellets to build the floor and the walls. Often, the overhang—or another nest—serves as the roof. Finally, they construct a narrow entrance. Because the swallows must wait for each layer of mud to dry before adding another, it takes a week or more to complete a nest.

Like miniature caves, the nests of cliff swallows ▷
face out from the side of a limestone cliff. Each nest
is made up of about 1,200 little balls of mud carried
there by a nesting pair of swallows. After drying,
the hardened nests may last for years.

△ *A cliff swallow uses its flat, broad bill as a shovel to*
scoop up mud for nest building. The bird holds up
its tail and long, pointed wings, preventing them
from getting wet and muddy.

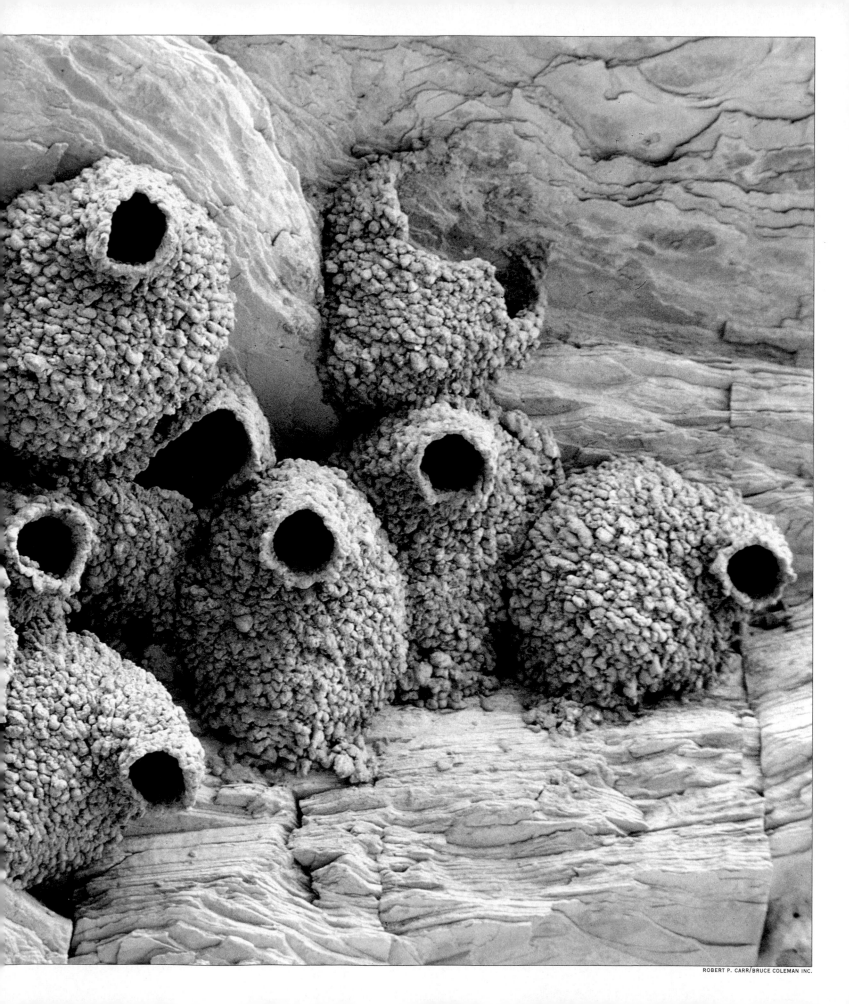

Ins and Outs of Bird Nests

Several kinds of birds besides swallows build nests of mud—but they use the mud differently. The black-browed albatross, for example, makes moundlike nests on steep slopes that face the sea. This large seabird nests in colonies on islands in southern oceans. To make its nest, an albatross pulls nearby mud to itself, piling it up and mixing it with grasses. The grasses strengthen the nest. The female shapes a shallow bowl at the top of the nest. Then she lays a single egg in it. The chick lives in the nest for four or five months after hatching.

In Africa, red-billed hornbills seek a more secret place to raise their young. When egg-laying time comes, the female selects a hole in a tree. Using mud brought to her by the male, she walls herself up inside the tree. She stays there for 45 days—laying up to five eggs, incubating them, and caring for the hatched young. Her mate provides the family with food. At first, he visits the nest about 30 times a day to bring fruit and insects for the female. Later, after the chicks hatch, he may make more than 70 trips a day. He feeds the family through a narrow slit in the mud seal.

When the chicks are about three weeks old, the female pecks her way out of the nest. The chicks patch the hole with their droppings, slugs, and sticky berries. They stay inside, fed through the slit by both parents. About three weeks later, they chip their way out and fly off.

G. ZIESLER/PETER ARNOLD, INC.

◁ *Overlooking the ocean, an adult black-browed albatross settles down to incubate its single egg. The moundlike nest, made of mud mixed with grasses, is often lined with soft feathers and grasses.*

A male red-billed hornbill, perched on the trunk of ▷ a hollow tree, holds a lump of moist soil in his huge bill. He will give the mud to the female, perched below him. She will use it to plaster up a hole in the tree. When just a crack remains, she will squeeze through the hole. Once inside, she will almost finish sealing it, leaving only a slit. The female will spend 45 days in this hideaway, laying and incubating eggs, and then caring for the young. The cutaway drawing (below) shows how the male feeds the female and nestlings through the slit. The mud seal helps keep out snakes and other enemies.

Insects That Move the Earth

At the edge of a puddle, you might come across a wasp molding a tiny ball of mud, beating its wings while it works. Chances are it will be a mud-dauber wasp collecting mud for its nest. The wasp will be a female, for among all wasps, only the females build nests.

There are many kinds of mud-dauber wasps, and they live in many parts of the world. They build their nests in barns and attics, under bridges, and in other sheltered places. A mud-dauber nest may be built in various shapes. Whatever its shape, the nest is made up of several compartments, called cells. Each cell is a nursery, complete with food, for a young wasp.

A female mud dauber builds a nest one cell at a time. When she finishes a cell, she goes hunting for spiders, which she paralyzes with her sting. She piles some spiders in the cell and lays an egg on one of them. After depositing a few more spiders, she seals up the cell with mud and begins a new cell. When she finishes the entire nest, she flies away—never to return.

When the eggs hatch, the wasp larvae feed on the spiders until the larvae are full grown. Then they spin cocoons inside the cells, changing into winged adults. The fully developed wasps chew their way out of their mud cells, one by one.

△ Can you see why the mud-dauber wasps that built these nests are called organ-pipe mud daubers? Each pipelike tube on the wooden beam has several cells. The holes show where wasps broke out of the cells after they hatched and developed into adults.

△ Using her legs and jaws, a female mud-dauber wasp scrapes up mud for her nest. She is careful to gather mud that is not too dry—and not too wet.

△ Inside a shed, a mud dauber seals up a cell. Each cell contains a wasp egg, along with food for the larva when it hatches. The most recently completed cell is still moist and therefore darker than the dried cells on either side of it.

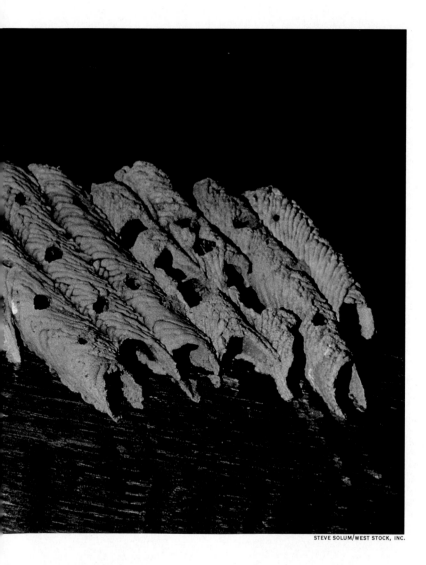

△ A mud nest with walls removed shows the developing wasps inside. At right are two cocoons with wasps in the pupal stage, before they become adults. At left is a pupa stripped of its cocoon.

Mud Makers

Unlike mud-dauber wasps, which gather existing mud, potter wasps and mason wasps make their own mud. First they fill their stomachs with water. Then they spit out drops of water to moisten dry soil. With the mud, they build some unusual structures.

The female potter wasp makes tiny mud-pot nests. She usually builds them on twigs of trees and shrubs or in cracks in rocks. The pots are about 1/2 inch (1 cm) wide. The wasp fills the bottom of each pot with paralyzed caterpillars or other insect larvae. Then she lays an egg and hangs it inside the pot by a silk thread. She plugs up the pot with mud. When the larva hatches, it finds itself well supplied with food.

A female mason wasp tunnels into sloping banks and digs out an underground nest. Inside, she builds nursery cells. With the mud that she removes to make the nest, the wasp builds a tube that curves out from the entrance. She lays an egg in each cell and stocks the cells with food for the larvae that will hatch. The wasp then uses mud from part of the tube to plug the nest entrance. This helps conceal the eggs and the food supply from animals that might find them tasty.

△ *Bit by bit, a mason wasp adds to the curved mud tube that leads to its nest. The wasp makes its own mud by mixing soil and water. The finished tube may be 6 inches (15 cm) long.*

◁ *A potter wasp completes a mud pot for use as a nest. Attached to a twig, the pot will hold one egg, and food for the wasp larva that hatches. The female will seal the pot with the egg and food inside.*

A female mason wasp has stung a beetle larva, ▷ *paralyzing it. Now she stuffs the larva into the entrance tube to her nest. Inside the nest, she will supply each mud cell with such a larva. These larvae will serve as meals for newly hatched wasps. The curved tube helps conceal the eggs and the food supply from enemies. It may help keep out rain, and help maintain an even temperature.*

Damp Domes and Chunky 'Chimneys'

Both of the animal architects on these pages work hard—one to avoid water, the other to reach it.

The soldier crab lives on sandy beaches in Southeast Asia. When the tide is low, the crab feeds, running about on the exposed sand. For safety, it digs an open burrow in the damp sand. It hides there at any sign of danger. When the tide comes in, an open burrow would collapse. Also, the crab needs a shelter because it would drown if it were underwater too long. So the crab builds a shelter with a roof. It starts running backward in a circle, pushing up pellets of sand that create a low, circular wall. As the crab continues around and around, the wall grows higher and higher, and finally curves inward. The crab then plugs the little hole at the top and is surrounded by an "igloo" of sand, complete with a pocket of air. The crab continues working, pushing sand from the floor to the ceiling of its igloo, deepening its chamber. Finally, in its shelter under the sand, the crab remains for several hours until the tide goes out.

Just as too much water is a threat to the soldier crab, too little water is a threat to the red swamp crayfish. It lives in swamps and other wet areas along the coast of the eastern and southern United States. Like other crayfish, it must keep its gills wet to breathe. This becomes difficult during the dry season, when ponds and swamps start to dry up. So the crayfish digs a well. It tunnels down to

IVAN POLUNIN/NHPA (BOTH)

△ A soldier crab begins to construct a dome-shaped shelter of wet sand as the tide comes in.

△ Within a short time, the crab has almost completed an igloo-like retreat. It will close up the hole.

△ To build its shelter, 1) the crab makes a shallow pit in the sand. 2) It runs around the pit backward, pushing up sand to form a curved wall. 3) The crab fills in the top with a pellet of sand, sealing in a pocket of air. 4) It burrows deeper beneath the thick dome of sand.

ART ADAPTED FROM "TWO CRABS OF THE SANDY SHORES," BY M. W. F. TWEEDIE, MALAYAN NATURE JOURNAL, VOL. 7, 1952

the water below ground level. Using its big front claws like bulldozer blades, it pushes mud out of the tunnel and piles it up around the entrance. As the crayfish digs deeper, the pile of mud grows taller. The pile begins to look like a chimney. If the crayfish tunnels deep enough, the chimney may become as tall as 2 feet (61 cm). The chimney hardens in the hot sun, preventing the crayfish from leaving its underground burrow for as long as four months. During this period, it never eats. It lives off the fat and muscle of its body. A female will lay eggs inside the tunnel. In the fall, rain comes and softens the mud chimney. When a female climbs out in the fall, she carries tiny hatched crayfish clinging to her legs.

▽ *Shoveling mud aside with its big front claws, a red swamp crayfish digs a burrow. During dry months, without surface water, the crayfish tunnels deeper and deeper to reach the water it needs for survival. A "chimney" forms around the burrow entrance from the mud that the crayfish removes.*

△ *Mud chimneys conceal the burrow entrances of red swamp crayfish. These chimneys are about as tall as pencils. The hot summer sun bakes the mud, making the chimneys almost as hard as brick. The crayfish stay inside their burrows until the fall. Then, when it rains, the chimneys soften and the crayfish climb out.*

5 Makers Of Wax And Paper

Building her nest, a queen *Polistes* wasp shapes thin-walled cells from bits of wood she has chewed into a moist pulp. When the wet pulp dries, the queen will have a sturdy paper nest for her young. Many kinds of wasps and bees make paper or wax for building complex structures.

N. A. CALLOW/NATURE PHOTOGRAPHER LTD.

Papering the House

In the spring, a queen *Polistes* (poh-LIH-steez) wasp hidden in bark, under a stone, or in another protected place wakes up from her winter's sleep. She selects a sheltered spot to build a nest, perhaps under a leafy branch or an overhanging roof, and begins to collect wood.

With her strong jaws, she scrapes bits of wood off telephone poles, fence posts, or fallen trees. She flies back to her building site with the bits of wood. There she chews the wood into a pulp, mixing it with her saliva. Shaping the moist pulp with her jaws, she forms a stem from which the nest will hang. To this she adds more pulp. She spreads it into thin layers and forms it into hollow cells. The pulp soon dries, hardening into tough walls of paper. The queen makes many trips to gather enough wood to build a nest of brood cells. In these cells, the young will be protected as they develop into adulthood.

Now she lays eggs, gluing one into each cell. About seven weeks later, new adults come out of the cells. These wasps, called workers, build more brood cells. The queen lays eggs in them. By summer's end, the expanded nest may have produced 200 or more wasps.

Like *Polistes* wasps, other paper wasps build nests for their young. They don't use the same nest design, however. Wasps called yellow jackets, for example, use abandoned burrows of mice and other creatures. Inside, they build nests that hang from the burrow roofs.

Little hornets, another kind of paper wasp, build nests of several stories. Each story contains hundreds of cells. Workers enclose the nest in a wall of paper several layers thick. The layers, and the air trapped between

◁ *About the size of a football, a paper wasp nest hangs from a branch. Wasps called little hornets enter the nest through a hole at the bottom. Workers regularly repair the outside of the nest.*

An opened nest of little hornets shows how the ▷ insects build an outer wall of paper several layers thick. The layers help insulate the nest from heat and cold. Young wasps—as pupae (PYEW-pea), in the white cocoons—require a constant temperature. (Never disturb a wasp nest! You could be stung.)

▽ *Snug as a bug, the larva of a bald-faced hornet lives in its paper brood cell. Adult hornets hunt insects to feed their helpless larvae.*

them, insulate the nest. This protects the young from cold and heat. After the egg and larval stages, the pupae (PYEW-pea) need a constant temperature while developing into adult wasps. If the temperature in the nest drops too low, workers create warmth by shivering together. If it gets too hot, they fan their wings in the nest to improve air circulation. If it's still too hot, workers carry drops of water into the nest and sprinkle them about.

Little hornets continually clean, repair, and enlarge their nest. To enlarge it, the workers chew the walls from the inside. This makes room for a new story of brood cells at the bottom of the nest. The wasps will add new paper layers on the outside.

Yellow jackets, a kind of paper wasp, add a layer of ▷ wet pulp to their nest. The variations in color are caused by the different kinds of wood the insects gathered to make the paper nest.

△ *To protect the nest from invading ants, Polistes (poh-LIH-steez) wasps smear a dark-colored ant repellent around the stem of the nest. Produced by glands in the wasps' abdomens, this chemical keeps away ants that might raid the nest to eat the young. The Polistes wasp has another effective weapon for keeping away unwanted guests—its stinger.*

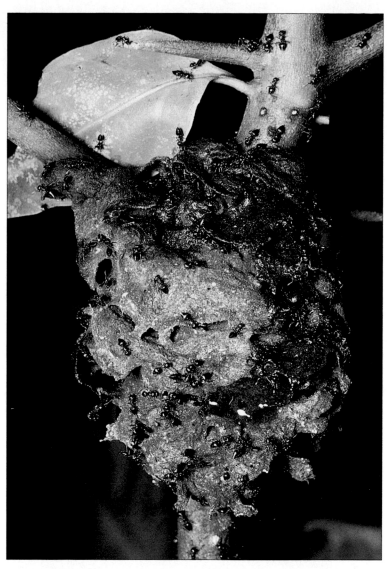

△ Built around a tree branch in West Africa, the paper nest of Crematogaster ants provides a cozy nursery for the developing young. The same basic papermaking process used by the insects was used to make the pages of this book. Wood was cut up into tiny chips and mixed into a pulp. After the pulp was pressed flat, it was allowed to dry.

Home, Sweet Home

Honeybees are master builders. By molding tiny pieces of wax, they construct a hive where as many as 30,000 bees live and work together.

The hive of the honeybee consists of honeycombs—wax walls with hundreds of small cells on each side. In the picture below, notice that all of the honeycomb cells are the same size. This perfect engineering is done by thousands of bees working together.

Bees use the cells as nurseries for the young and as containers for storing food. Workers collect pollen and nectar from flowers. They use the nectar—which they make into honey—and the pollen to feed the colony.

The bees' six-sided cell is a useful design. The cells fit together, sharing walls. This provides the most storage space with the least wax. Each cell wall is thinner than a page of this book. Yet the cells are so sturdy, they can support many times their own weight.

As the population of a colony grows, worker bees produce wax and add new honeycombs. Glands in their abdomens produce a clear liquid. The liquid oozes from the glands and hardens into white flakes of wax. The bees pick up the wax with small hooks on their legs. They put the wax into their mouths and work it around in their jaws until it is soft.

Workers then team up with hundreds of other wax-making bees. They use their jaws to apply the wax and shape it into honeycombs. Heat generated by so many bees moving together keeps the work space warm enough so that the wax remains soft and workable. Each bee labors for about 30 seconds before another takes its place, picking up where the last bee left off. The bees work steadily until they have built enough cells for the colony. They are "as busy as bees"!

After a colony has grown in size, part of it may leave in a swarm—a mass of bees. Scouts look for a place for a new hive. In warm climates, honeybees may choose the leafy shade of a tree branch. In cooler climates, they usually select a spot protected from wind and cold—in the hollow of a tree, for example.

Honeybees frequently move into hives supplied by

◁ Honeybees bunch together in their hive. A hive is made up of wax walls called honeycombs. Each side of a honeycomb may contain hundreds of wax cells used for raising young or for storing food.

JEFF FOOTT

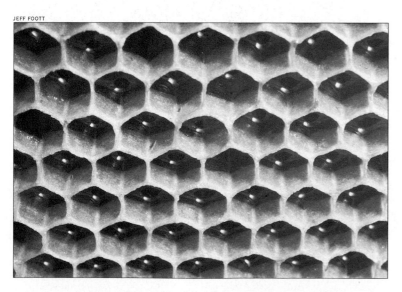

△ A closeup view of a honeycomb shows wax cells filled with honey. Bees make honey from nectar, a sweet liquid they gather from flowers.

JOHN MASON/ARDEA LONDON

△ Fast-growing honeybee larvae live in wax brood cells after hatching there from eggs. Worker bees will feed the larvae for about six days. The larvae then will change into pupae, and grow to be adults. Worker bees will clean the empty brood cells to prepare for the next batch of eggs.

beekeepers. These wooden boxes have vertical frames on which the bees build their honeycombs. Beekeepers regularly harvest the honey and beeswax. They make sure, however, to leave enough honey for the bees to eat through the winter. The beeswax is used to make products such as crayons, gum, candles, and cosmetics.

Unlike wasps, which in northern climates occupy their nest for only one season, honeybees may live in a hive for many years. In winter, the bees cluster together for warmth. They open the wax lids of the food-storage cells and feed on the honey inside. When spring arrives, the colony becomes active again. The bees begin to gather and store food, and to build new honeycombs.

Worker honeybees eat honey from open storage ▷ cells. Cells at the upper left have wax lids that will preserve the honey until winter. Capped cells at the lower right hold pupae. The pupae lie inside cocoons they have spun. They were sealed in the brood cells by worker bees. When the pupae are grown, they will break out of the cells as worker bees.

△ *A honeybee pauses, while buzzing from flower to flower, to collect nectar and pollen. It sips the nectar with its long tongue and carries the powdery yellow pollen on the stiff hairs of its hind legs.*

▽ A queen bumblebee warms a cluster of wax brood cells that hold her eggs. Bumblebees, in groups of 100 to 400, make small nests among clumps of grass or in abandoned mouse burrows. The queen keeps a wax pot of honey nearby for meals.

▽ A worker bumblebee drinks honey from a pot. Honeypots supply the colony with food during cold or rainy weather. To make honey, bumblebees collect nectar, adding a chemical as they work. Over time, the nectar changes into honey.

6 Excavators: Animals That Dig

Fresh dirt flies as a black-tailed prairie dog clears an entrance to its underground burrow. The rodent digs with its front feet and kicks out dirt with its hind feet. Other diggers you'll read about build sand pits, dens in the snow, and stone-lined burrows in the ocean floor.

© JERRY L. FERRARA

Burrow Builders

In the mountains of North America, Europe, and Asia where marmots live, wintry cold may last most of the year. On the Great Plains of North America—home of the black-tailed prairie dog—the land is baked by summer sun and whipped by winter snow.

To escape the harsh temperatures, marmots and prairie dogs dig burrows. There, the soil and air temperatures stay comfortable year-round. Narrow tunnels wind down 15 to 20 feet (4½ to 6 m) to well-insulated chambers. Opening off the tunnels are chambers for sleeping, for raising young, and even for use as toilet areas.

Marmot burrows are much simpler than prairie dog burrows. Marmots usually use one entrance that leads to a few large chambers. The whole family—as many as 15—sleeps in this burrow. Rarely does the burrow connect with other marmot burrows. In contrast, groups of black-tailed prairie dogs work together to dig complicated burrow systems with many openings, tunnels, and chambers. The systems may connect with each other.

A marmot removes a stone from its underground ▷ home in a mountain meadow in Washington State. It shares its burrow with other family members. In the summer, a marmot may dig other, temporary burrows in feeding areas to escape enemies.

△ *With a quick touch, a black-tailed prairie dog and its young identify each other. The mound around their burrow hole helps keep water from flowing in.*

86

Down the Rabbit Hole

For rabbits, life above the ground is dangerous. Enemies such as foxes, dogs, hawks, and owls prey on them. Large colonies of European rabbits protect themselves by living underground. The rabbits build complex systems of underground burrows called warrens. A single warren has many entrance holes and may occupy 2½ acres. In England, one study of a colony of 400 rabbits found more than 2,000 entrances to the warren.

European rabbits dig their warrens in sandy soil in fields or at the edges of wooded areas. Main entrance holes are marked by dirt mounds worn down by the feet of rabbits entering or leaving the warren. Smaller holes, called bolt holes, are well hidden by vegetation. The rabbits use them to bolt, or run, from enemies if caught feeding in the open. If you could climb down into a warren, you would find a maze of tunnels and many sleeping chambers. Burrowing rabbits sometimes come upon large rocks or roots. As a result, a tunnel may twist and turn, branch into two passages, or come to a dead end.

Most females about to give birth build nesting burrows separate from the warren. They line these burrows with dry plants and with fur plucked from their bellies.

During the day, European rabbits rest in their warren. At dusk, during the night, and at dawn, they come out to feed. They take care, though, not to go too far from the safety of their underground home.

▽ *European rabbits dig a system of underground burrows called a warren. At dawn, a warren is a busy place, as this cutaway painting shows. In the open, rabbits feed while listening for danger. Others come and go through scattered entrances. Belowground, a rabbit rests in a sleeping chamber. Another scurries along a tunnel. One rabbit digs a new tunnel. A female nestles with her young in a nesting burrow lined with dry plants and fur. Plants conceal the entrance to this burrow.*

A European rabbit comes out of its burrow ▷ entrance. To dig, a rabbit scratches in sandy soil with its front paws and tosses dirt with its hind legs.

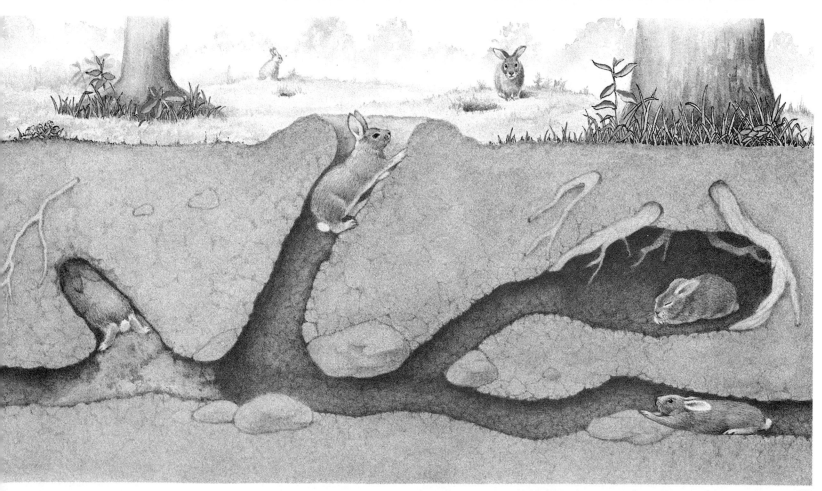

Like prairie dogs, marmots, and European rabbits, the animals on these pages—foxes and North American badgers—dig burrows. But badgers and foxes use burrows differently from the other animals. Badgers may change burrows daily. Some foxes use theirs only part of the year.

The North American badger is most at home on the open prairie. Except for a female with cubs, the badger lives alone. It hunts at night for ground squirrels, rabbits, and other small mammals.

The badger's long front claws and powerful legs make it one of the world's best diggers. It can dig itself completely underground within a few minutes. The badger digs to unearth small animals to eat. It digs burrows for sleeping and to avoid bad weather. It digs itself out of sight to escape enemies.

A North American badger's burrow consists of a long tunnel with a chamber for sleeping. In the summer, the badger often digs a new burrow every day. One badger was found to have dug 50 burrows in its feeding area. In the winter, the badger may stay inside one burrow for long periods—occasionally coming out to hunt. It lines its burrow with vegetation, making a soft resting area. In the spring, a female digs a nursery burrow before giving birth. There, she'll nurse her cubs for six to eight weeks.

Sometimes a badger takes over a burrow dug by another animal and changes it to suit its needs. In a similar way, an abandoned badger burrow often provides shelter for rabbits and other animals.

Unlike the badger, the red fox prefers to let other animals do its digging. In late winter, the female fox looks for an empty burrow to use as a den for raising her young. If necessary, she digs a den. Inside, a tunnel—sometimes 30 feet (9 m) or longer—leads to a chamber. There, she gives birth. She nurses the kits while the male hunts and brings food to her. As the kits grow older, they use the den less and less—mainly for hiding from enemies. In the fall, when the kits have learned to hunt for themselves, the family leaves the den. Each fox goes off on its own.

TOM AND PAT LEESON

△ *A young North American badger digs in the loose dirt around the entrance to its burrow. For this youngster, digging is like a game. But the animal is developing a skill it will need for survival. Adult badgers spend much of each day digging. They dig to make burrows, to uncover food such as mice and prairie dogs, and to escape danger.*

▽ *With its job done, a badger crawls from a freshly dug burrow. It often sleeps in the burrow for just one day, then digs a new one the next. Long, sharp claws help the badger dig quickly.*

W. PERRY CONWAY/TOM STACK & ASSOCIATES

STEPHEN J. KRASEMANN/DRK PHOTO

△ In East Africa, a bat-eared fox stands at the entrance to one of its several burrows. By day, it uses the burrows to avoid the heat and to hide from enemies. By night, it digs for termites to eat. It can hear them underground with its huge ears.

▽ On a spring morning, two red fox kits peek from their underground den. Their mother either dug the den herself, or took over and enlarged a burrow abandoned perhaps by a marmot or a badger. A red fox den may have as many as ten openings.

LEONARD LEE RUE III/LEN RUE ENTERPRISES

Snowy Den

Polar bear cubs are born at the coldest time of year in one of the coldest regions on earth—the Arctic. The tiny cubs are helpless at birth. How do they survive the extreme cold? They spend the first few months cuddled against their mother in a snug snow den. The cubs get energy from her rich milk.

In the fall, a pregnant polar bear digs a den in drifted snow. With swipes of her huge paws, she scoops out a living space. She may punch a hole in the ceiling to let in fresh air. If you were nearby, you'd see the bear's breath rising out of the ventilation hole like smoke.

In December, the female gives birth, usually to two cubs. The cubs nurse inside the den for three or four months, growing thick fur coats and developing a layer of fat that will protect them from the cold. The temperature in the den may be 40°F (22°C) warmer than the temperature outside. In the spring, the female tunnels through the snow out of the den. The cubs stumble out after her and see bright sunshine for the first time.

◁ A fresh hole shows where a female polar bear and her cubs broke out of their winter snow den. Inside, one or two rooms sheltered the bears. The female gave birth in the den in December. She remained inside with her cubs for several months. The den protected them from the winter winds, while the mother's body warmed the cubs as they nursed and grew warm fur coats. For a few days, the bears return to this den from time to time to rest.

JACK W. LENTFER

△ A female polar bear in Alaska sniffs the spring air
for the scent of prey. Polar bears mate in the spring.
In the fall, a pregnant female looks for a mound of
drifted snow. Digging in, she hollows out a den,
where her cubs will be born. Drifting snow closes
the entrance, helping protect her from the wind and
cold. Males and nonpregnant female polar bears
remain active throughout the winter.

In the fall, a polar bear nurses her cubs, now nearly ▷
a year old. The three will stay together through the
approaching winter—and the next. With thick fur,
and with protective fat under their skin, the cubs
won't need a permanent den in the coming winter.

BRYAN AND CHERRY ALEXANDER

93

'Holesome' Birdlife

Knock-knock-knock-knock! That's the sound of one of the world's largest excavating birds. You can hear the hammering of the pileated woodpecker in many North American forests. If you should hear it, you might think you were listening to a carpenter at work.

In a way, the pileated woodpecker is a carpenter. Using its chisel-like bill, the crow-size bird pecks away at tree trunks in the spring to make holes for nesting. Perched on the side of a tree, it grips the bark with its long, curved claws. Stiff tail feathers prop it up. The woodpecker's thick skull, and space between the skull and the brain, protect the brain from the battering.

Chiseling out the nest cavity is hard work. It may take as long as a month. A completed nest cavity measures about 2 feet (61 cm) deep and 6 to 8 inches (15 to 20 cm) wide. Wood chips line the bottom of the nest hole. On the bed of chips, the female lays her eggs. The male and female take turns incubating the eggs and feeding the young woodpeckers.

△ *Having fed its chicks, a sand martin clings to the edge of its nest hole, in England. Both parents use their bills and feet to tunnel as deep as 4 feet (1¼ m) into a sandbank. They make a nest at the end.*

△ *Thrusting out its neck, a male pileated (PYE-lee-ate-uhd or PILL-ee-ate-uhd) woodpecker peers from the nest hole he and his mate carved in a dead tree. They used their chisel-like bills to hack out the cavity. Together they will raise three to five young.*

△ An acorn woodpecker perches on a dead tree that's as full of holes as Swiss cheese. The woodpecker drills holes to store acorns, one of its main winter foods. A group of two to ten acorn woodpeckers prepares and defends such a storage site. One tree may contain as many as 50,000 holes.

Acorns fit neatly into holes drilled in a post. In the ▷ fall, acorn woodpeckers gather the acorns and use their bills to tap them into the holes. In winter, when other food is scarce, the birds eat the acorns.

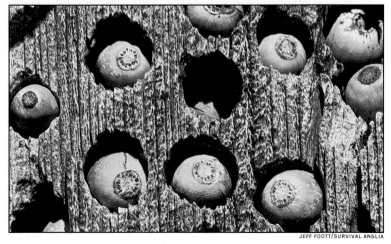

Diggers of the Sea

Each of the two very different fish shown on these pages digs a burrow for itself. You might wonder how the fish do this without any arms or legs. Read on to find out.

The mudskipper is one of the few fish that can survive on land. It lives along the coasts of Africa and Southeast Asia in burrows it builds at low tide by scooping up mud with its mouth. At high tide, the fish lives underwater in its mud burrow. But when the tide goes out, the fish stays behind. It leaves its burrow and drags itself around on the mud with its fins. It breathes on land by keeping sacs around its gills filled with water and air.

When ready to breed, the male performs a courtship display. He attracts a mate by flipping about on the mud and dancing on his stubby fins. In a water-filled burrow that may be as deep as 2 feet (61 cm), the female lays eggs and the male fertilizes them. One of the pair always guards the eggs in the burrow.

The yellowhead jawfish, in contrast, is an underwater architect. It builds its burrow near coral reefs on the ocean floor from Florida to South America. The jawfish uses its burrow to hide from larger fish that prey on it.

To construct its burrow, the yellowhead jawfish opens its large jaws and scoops up mouthfuls of sand. It spits out the sand to the sides, continuing this activity until it has a cone-shaped pit. The fish then begins to fill the pit with rubble—bits of coral and shell and small pieces of rock. It presses the rubble into place. This prevents the sand from caving in as the jawfish digs a chamber below the pit. Within the rubble, the fish leaves a tunnel to the chamber.

It's a close fit for the jawfish in its tunnel. Whenever the fish senses danger, it scoots in, usually tailfirst. At night, the fish covers the tunnel opening from the inside by placing a large piece of rubble over it.

S. CORDIER/PITCH

△ Along the coast of West Africa, a mudskipper guards its burrow on a mud flat. The fish uses the burrow for resting at high tide, and for breeding. By keeping its gills moist, a mudskipper can breathe on land. It uses its armlike fins to move around.

*Ready to gobble up tiny animals that ▷
float by, a yellowhead jawfish in an
aquarium hovers over the entrance
to the burrow it has constructed.*

*▽ A jawfish retreats tailfirst into its
burrow to escape from larger fish
or simply to rest.*

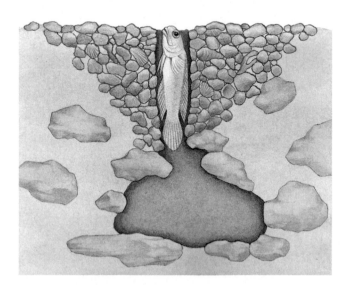

*△ The jawfish just fits in the tunnel of its
burrow. The fish digs the burrow in the
seafloor using its mouth. It lines the top
with coral, shell, and rock, leaving a
tunnel to the chamber underneath.*

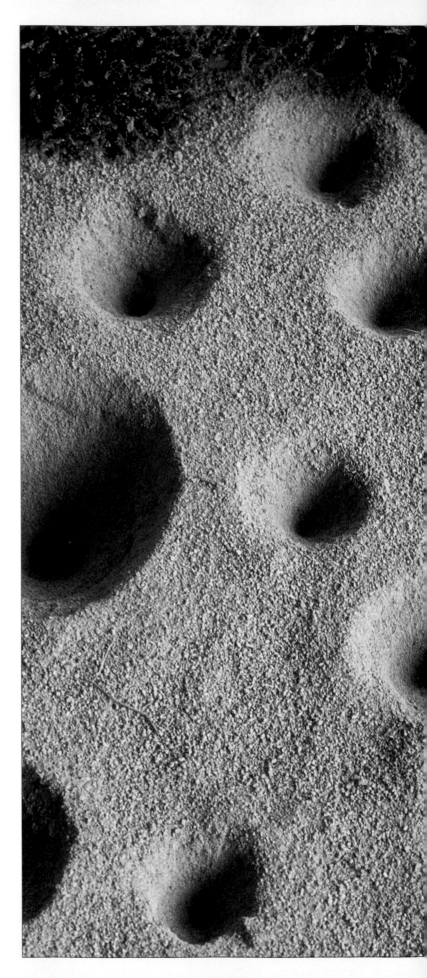

Sand Traps

An ant lion is neither ant nor lion. It's an insect named for the behavior of its larvae. An ant lion larva preys mainly on ants, and, like a lion, ambushes its prey.

There are more than 600 kinds of ant lions. They live in warm regions all over the world. Many larvae trap their prey in small pits they dig in loose, dry soil. To construct a pit, an ant lion circles backward in ever smaller circles. As grains of sand or soil fall on its head, it tosses them away. Eventually, it makes a 2-inch-deep (5-cm) crater-shaped pit. At the bottom of the pit, the larva pushes its abdomen into the dirt and digs itself in. Only its jaws remain exposed. Once buried, the ant lion waits.

When an ant or other insect comes too close to the edge of the pit, it slides in. It may fall to the bottom, where the ant lion snaps it up with its jaws. If the insect struggles to climb out of the pit, the ant lion tosses up pieces of soil or sand. They may hit the insect or start a small slide. In either case, the insect usually comes tumbling down to the ant lion's jaws. Like a spider, the ant lion holds its prey and sucks out the body fluids.

Buried at the bottom of each of these pits, an insect ▷ larva called an ant lion lies in wait for another insect to fall in. Ant lions circle backward and toss dirt with their heads to dig the craterlike traps.

△ *An ant lion digs into the sand backward at the bottom of its pit. It will bury itself so that only its jaws remain uncovered—ready to seize prey that falls in. Bristly hairs on its body will help keep the larva in place when it holds on to struggling prey.*

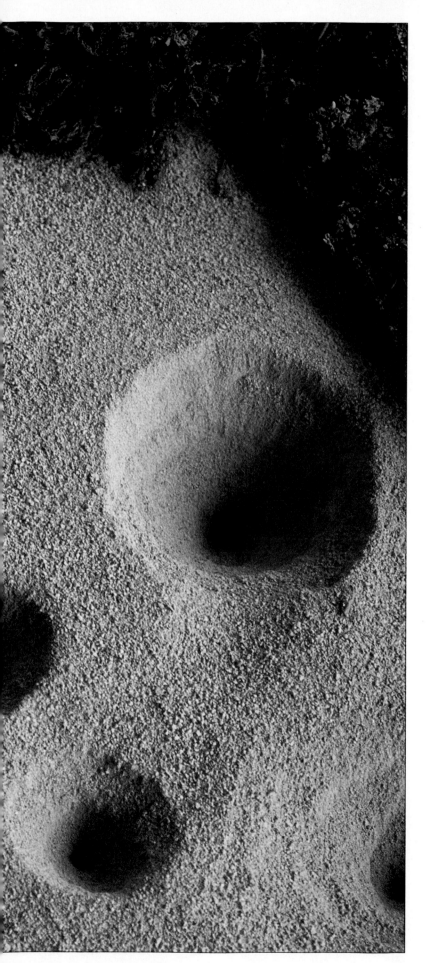

ROBERT L. DUNN/BRUCE COLEMAN INC.

△ *With its sawtooth jaws, an ant lion reaches for an ant in its trap. An ant lion usually remains in the larval stage for two years, feeding on insects.*

HANS PFLETSCHINGER/PETER ARNOLD, INC. (BELOW AND LEFT)

△ *Inside a sand-covered cocoon, an ant lion pupa develops into its adult form. As a larva, the insect spun the silk cocoon while buried in its pit. Sand stuck to the outside. (The cocoon was uncovered for the picture.) After about a month, the adult will climb out, let its wings dry, and fly off.*

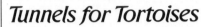

Tunnels for Tortoises

The desert tortoise spends as much as 90 percent of its life underground. It lives in dry areas of the southwestern United States and northern Mexico. There, in summer, the sun blazes, making temperatures soar. In winter, temperatures sometimes fall below freezing. The tortoise adapts to these extremes by digging two types of summer burrows, in which it lives alone. It also digs a winter burrow, which it may share with other desert tortoises.

In warm weather, the tortoise excavates shallow burrows in sand or gravel, often under shrubs. Scraping at the ground with its claws and broad front legs, it tunnels in. When the dirt piles up behind it, the tortoise turns around. Pushing forward with its hind legs, it shoves the dirt out of the hole, using its body like a bulldozer.

The completed burrow is the right size and shape for one desert tortoise. The tortoise stays inside this burrow

during the heat of the day. In the early morning and in the evening, when the air is cooler, the tortoise trudges out to eat grasses, leaves, cactuses, and blossoms. In the hottest part of summer, the tortoise moves to a deeper burrow. There it sleeps for days or even weeks.

To stay warm in winter, desert tortoises build much larger, deeper burrows, called dens. The main entrance tunnel to a den may extend more than 30 feet (9 m) into a gravel slope—far enough to avoid freezing temperatures. In a central chamber, as many as a dozen desert tortoises may hibernate. Like the other creatures in this book, desert tortoises are masters of animal architecture.

▽ *One desert tortoise meets another outside a warm-weather burrow. These desert architects dig shallow summer burrows where they live alone. The visitor will have to dig its own.*

INDEX

ADDITIONAL READING

Readers may want to check the *National Geographic Index* and the *National Geographic World Index* in a school library or a public library for related articles.

These volumes in the National Geographic Society's Books for World Explorers series also contain related material: *How Animals Behave: A New Look at Wildlife; Secrets of Animal Survival; The Secret World of Animals;* and *Wildlife Alert! The Struggle to Survive.* Readers may also want to refer to the Society's *National Geographic Book of Mammals, Vols. 1 and 2;* and *Wild Animals of North America,* as well as to the following books ("A" indicates a book written for adults):

Bare, Colleen Stanley, *The Durable Desert Tortoise,* Dodd, Mead & Co., 1979. Hancocks, David, *Master Builders of the Animal World,* Hugh Evelyn Ltd., 1973 (A). Hansell, M.H., *Animal Architecture*

Day comes to an end for a pair of white storks on their nest in Morocco, a country in North Africa. The birds will add to the pile of sticks throughout the nesting season. The nest may measure as much as 6 feet (2 m) across. Storks build their nests atop walls, towers, chimneys, roofs, and platforms, and in trees. People like having the birds around. Storks are thought to bring good luck, and they eat small animals such as grasshoppers and mice, which farmers don't want to have around.

and Building Behaviour, Longman Group Ltd., 1984 (A). Lavine, Sigmund A., *Wonders of Badgers*, Dodd, Mead & Co., 1985. Norsgaard, E.J., *Insect Communities*, Grosset & Dunlap, 1973. Nussbaum, Hedda, *Animals Build Amazing Homes*, Random House, Inc., 1979. Van Wormer, Joe, *Squirrels*, E.P. Dutton, 1978. Von Frisch, Karl, *Animal Architecture*, Harcourt Brace Jovanovich, 1974 (A). Warner, G.F., *The Biology of Crabs*, Van Nostrand Reinhold Co., 1977 (A).

CONSULTANTS

Fiona Sunquist, *Chief Consultant*

Thomas A. Jenssen, Ph.D., Virginia Polytechnic Institute and State University, *Consulting Herpetologist*

William J. B. Miller, Florida Department of Natural Resources, Florida Park Service, *Consulting Ichthyologist*

Walter S. Sheppard, Ph.D., U.S.D.A., Agricultural Research Service, Beneficial Insects Laboratory, *Consulting Entomologist*

George E. Watson, Ph.D., St. Albans School, Washington, D.C., *Consulting Ornithologist*

Glenn O. Blough, LL.D., Emeritus Professor of Education, University of Maryland, *Educational Consultant*

Nicholas J. Long, Ph.D., *Consulting Psychologist*

Barbara J. Wood, M.Ed., Montgomery County (Maryland) Public Schools, *Reading Consultant*

The Special Publications and School Services Division is also grateful to the individuals and institutions named or quoted in the text and to those cited here for their generous assistance:

Benjamin Beck, National Zoological Park, Smithsonian Institution; Larry de la Bretonne, Jr., Louisiana State University; Alison Brooks, George Washington University; Valerie Chase, National Aquarium in Baltimore; Nicholas Collias, University of California at Los Angeles; Charles J. Farwell, Monterey Bay Aquarium; Gregory L. Florant, Swarthmore College.

Jack W. Lentfer, Environmental Research and Consulting; Charles H. Lowe, University of Arizona; Gail R. Michener, University of Lethbridge; Edward O. Murdy, Smithsonian Institution; Duane A. Schlitter and Steve Williams, Carnegie Museum of Natural History; Victor G. Springer, Smithsonian Institution; Austin Williams, National Marine Fisheries Service.

Composition for ANIMAL ARCHITECTS by the Typographic section of National Geographic Production Services, Pre-Press Division. Printed and bound by Holladay-Tyler Printing Corp., Rockville, Md. Film preparation by Catharine Cooke Studio, Inc., New York, N.Y. Color separations by Lincoln Graphics, Inc., Cherry Hill, N.J.; and NEC, Inc., Nashville, Tenn. Teacher's Guide printed by McCollum Press, Inc., Rockville, Md.

Library of Congress CIP Data

Animal architects
(Books for world explorers)
Bibliography: p.
Includes index.
Summary: Discusses the structures built by many different kinds of animals for protection, food gathering, storage, and nesting.
1. Animals—Habitations—Juvenile literature. [1. Animals—Habitations]
I. National Geographic Society (U. S.) II. Series.
QC756.A53   1987   591.56'4   87-12198
ISBN 0-87044-612-6 (regular edition)
ISBN 0-87044-617-7 (library edition)

ANIMAL ARCHITECTS

PUBLISHED BY
THE NATIONAL GEOGRAPHIC SOCIETY
WASHINGTON, D. C.

Gilbert M. Grosvenor, *President and Chairman of the Board*
Melvin M. Payne, *Chairman Emeritus*
Owen R. Anderson, *Executive Vice President*
Robert L. Breeden, *Senior Vice President, Publications and Educational Media*

PREPARED BY THE SPECIAL PUBLICATIONS
AND SCHOOL SERVICES DIVISION

Donald J. Crump, *Director*
Philip B. Silcott, *Associate Director*
Bonnie S. Lawrence, *Assistant Director*

BOOKS FOR WORLD EXPLORERS
Pat Robbins, *Editor*
Ralph Gray, *Editor Emeritus*
Ursula Perrin Vosseler, *Art Director*
Margaret McKelway, *Associate Editor*
David P. Johnson, *Illustrations Editor*

STAFF FOR *ANIMAL ARCHITECTS*
Sharon L. Barry (Chapters 4, 6), Susan McGrath (Chapters 1–3), Suzanne Venino (Chapter 5), *Writers*
Roger B. Hirschland, *Managing Editor*
Veronica J. Morrison, *Picture Editor*
Louise Ponsford, *Art Director*
Debra A. Antonini, *Senior Researcher*
Catherine D. Hughes, Bruce G. Norfleet, Suzanne Nave Patrick, *Contributing Researchers*
Patricia N. Holland, *Special Projects Editor*
Biruta Akerbergs Hansen, *Artist*
Roz Schanzer, *Artist (Headline art)*
Joan Hurst, *Editorial Assistant*
Bernadette L. Grigonis, Artemis S. Lampathakis, Karen L. O'Brien, *Illustrations Assistants*
Aimée L. Clause, *Art Secretary*
Catherine H. Phillips, Terry Stutzman, *Editorial Interns*

ENGRAVING, PRINTING, AND PRODUCT MANUFACTURE: Robert W. Messer, *Manager*; George V. White, *Assistant Manager*; David V. Showers, *Production Manager*; Timothy H. Ewing, *Production Project Manager*; Gregory Storer, George J. Zeller, Jr., *Senior Assistant Production Managers*; Mark R. Dunlevy, *Assistant Production Manager*; Carol R. Curtis, *Production Staff Assistant*.

STAFF ASSISTANTS: Katherine R. Davenport, Mary Elizabeth Ellison, Donna L. Hall, Mary Elizabeth House, Bridget A. Johnson, Sandra F. Lotterman, Eliza C. Morton, Nancy J. White, Virginia A. Williams.

MARKET RESEARCH: Mark W. Brown, Joseph S. Fowler, Carrla L. Holmes, Marla Lewis, Barbara G. Steinwurtzel, Marsha Sussman, Lisa A. Tunick, Judy Turnbull.

INDEX: Susan G. Zenel